Silent Voices From the Past

SILENT VOICES

from the Past

A Chronicle of the Almshouse of Sullivan County

SARA POISSON

iUniverse, Inc.
New York   Bloomington

Silent Voices From the Past
A Chronicle of the Almshouse of Sullivan County

iUniverse books may be ordered through booksellers or by contacting:

iUniverse
1663 Liberty Drive
Bloomington, IN 47403
www.iuniverse.com
1-800-Authors (1-800-288-4677)

ISBN: 978-1-4502-5907-1 (sc)
ISBN: 978-1-4502-5908-8 (ebk)

Printed in the United States of America

iUniverse rev. date: 09/15/2010

This book is dedicated to:

My mother, Marion McIntyre for instilling appreciation of and respect for cemeteries

My husband, Paul for putting up with it

Special thanks to:

The men of the Transitional Housing Unit for hours of research

The Newport Historical Society without whom this would not have been possible

Introduction

The idea for this book came to me one day as I walked from my office in the old part of the Sullivan County Nursing Home to the jail where I work as a mental health and substance abuse clinician. The walk takes me near a cemetery where one can easily see a stark difference between the front half of the cemetery which is ornate and obviously very old. The back half of the cemetery has very small uniform stones that are in perfect rows, not in family plots as one would expect to see. Upon investigation, I learned that these small nondescript stones each belongs to someone who died here on this land during the time period that this spot was the County Almshouse. The look of the stones, the effective use of space and the lack luster appearance of this half of the cemetery is indicative of how these folks were looked upon. These were the throwaways of society, the ones no one wanted to see or know about. It struck me that there is a strong parallel between these rows of folks and those that I work with.

These folks are also the throwaways of society. They also have stigma. They also have a difficult time in the world financially and socially because of the class of people they belong to.

Part of my job is to help these folks understand that they do have worth and that they can and should make positive contributions to society through successfully living their own lives. Unfortunately, most of these folks are not proud of who the have become. Seeing the worth in oneself is difficult when it is hard for others to do so. It occurred to me that this was an opportunity to teach self-respect, hard work, compassion and respect for others through a research project. This book is the result of that work. Each and every person in this part of the cemetery has been identified. All that could be learned about each of these persons is recorded here. Additionally, the history of the evolving cultural and institutional norms is here as well giving all the information about building and farming improvements that we could find.

I hope that some of the twenty or so young men that worked on this project have learned that their lives have meaning, that they can and should make a difference and that there should be equal respect for all people regardless of personal history, economic status, age or general misfortune.

SULLIVAN COUNTY FARM BUILDINGS.

The purpose of the poorhouse was to provide food and lodging for the indigent of the towns and county. The poor house population generally exploded in the early 1800's where conditions ranged from poor to horrific. Often times the leadership was won by political favor and as little as possible was done to earn a paycheck. Everything possible was done to keep the cost down and the population low. The first entry in the Poor Farm registry was dated December 16, 1867. The person's name was Sybil Kindal. She as a single white female from Claremont who was born in Swanzey, New Hampshire. The last entry is for Mr. James Young, a 55 year old white single male from Claremont who was born in Laconia. He was admitted September 23, 1931. Outdoor relief was similar to today's welfare system. Money was given to the poor to use for their living expenses. Indoor relief refers to the Almshouse system. This work covers the beginning of the Almshouse in 1867 until 1938 when the Almshouse system had become archaic.

Ending up in the poor house was a shameful and depression situation in life. In 1897 Will Carleton wrote "Over the Hill to the Poor-House" the lyrics of which follow:

Over the hill to the poor-house I'm trudgin' my weary way..
I, a woman of seventy, and only a trifle gray..
I, who am smart an' chipper, for all the years I've told,
As many another woman that's only half as old.

Over the hill to the poor-house—I can't quite make it clear!
Over the hill to the poor-house—it seems so horrid queer!
Many a step I've taken, a –toilin' to and fro,
But this is the sort of journey I never thought to go.

What is the use of heapin' on me a pauper's shame?
Am I lazy or crazy? am I blind or lame?
True, I am not so supple, nor yet so awful stout;
But charity ain't no favor, if one can live without.

I am ready and willin' an' anxious any day
To work for a decent livin' and pay my honest way;
For I can earn my victuals, an' more too, I'll be bound,
If anybody is willin' to only have me 'round.

Once I was young an' handsome—I was, upon my soul—
Once my cheeks was roses, my eyes as black as coal;
And I can't remember, in them days, of hearin' people say,
For any kind of a reason, that I was I their way!

T'aint no use of boastin' or talkin' over-free,
But many a house an' home was open then to me;
Many a han'some offer I had from likely men,
And nobody ever hinted that I was a burden then.

And when to John I was married, sure he was good and smart,
But he and all the neighbors would own I done my part;
For life was all before me, an' I was young an' strong,
And I worked my best an' smartest in tryin' to get along.

And so we worked together; and life was hard, but gay,
And now and then a baby to cheer us on our way.
Till we has half a dozen, an" all growed clean an' neat,
An" went to school like others, an' had enough to eat

An' so we worked for the childr'n, and raised 'em every one—
Worked for 'em summer and winter, just as we ought
    to've done.
Only perhaps we humored 'em, which some good folks
    condemn,
But every couple's own child'rn's a heap the dearest to them!

Strange how much we think of OUR blessed little ones!—
I'd have died for my daughters, and I'd have died for my sons.
And God He made the rule of love; but when we're old
    and gray
I've noticed it sometimes, somehow, fails to work the
    other way.

Stranger another thing; when our boys an' girls was grown,
And when, exceptin' Charley, they'd left us there alone,
When John he nearer and nearer came, an' dearer seemed
    to be,
The Lord of Hosts, He came one day an' took him away
    from me!

Still I was bound to struggle, an' never cringe or fall—
Still I worked for Charley, for Charley was now my all;
And Charley was pretty good to me, with scarce a word
    or frown,
Till at last he went a courtin' and brought a wife from town.

She was somewhat dressy, an' hadn't a pleasant smile—
She was quite conceity, and carried a heap o' style;
But if ever I tried to be friends, I did with her, I know;
But she was hard and haughty, an' we couldn't make it go.

She has an edication and that was good for her,
But when she twitted me on mine, 'twas carryin' things
    too far,
An' I told her once, 'fore company, (an' it almost made
    her sick)
That I never swallowed a grammer, nor 'et a 'rithmetic.

So 'twas only a few days before the thing was done—
They was a family of themselves, and I another one.
And a very little cottage one family will do,
But I have never seen a mansion that was big enough for two.

An' I never could speak to suit her, never could please
    her eye,
An' it made me independent, an' then I didn't try.
Bit I was terribly humbled, an' felt it like a blow,
When Charley turned agin me, an' told me I could go!

I went to live with Susan, but Susan's house was small,
And she was always a hintin' how snug it was for us all;
And what with her husband's sisters, and what with
    child'rn three,
"Twas easy to discover there wasn't room for me.

An' then I went to Thomas, the oldest son I've got;
For Thomas's buildings'd cover the half of an acre lot,
But all the child'rn was on me and I couldn't stand their sauce
And Thomas said I needn't think I was comin' there to boss.

An' then I wrote to Rebecca, my girl who lives out West,
And to Issac, not far from her—some twenty miles at best;
And one of 'em said 'twas too warm there for anyone so old,
And t'other had an opinion the climate was too cold.

So they have shirked and slighted me, an' shifted me about—
So they have well nigh soured me, an' wore my old hear out;
But still I've borne up pretty well, an' wasn't much put down,
Till Charley went to the poor-master, an' put me on the town!

Over the hill to the poor-house—my child'rn dear good-bye!
Many a night I've wateched you when only God was nigh;
And God'll judge between us; but I will al'ays pray
That you shall never suffer the half that I do to-day!

The Department of Commerce and Labor, Bureau of
the Census S.N.D. North, Director in his special report
Paupers in Almshouses (1904) reported that the law in
New Hampshire read:

The overseers of the poor in any town must relieve and
maintain the poor of such towns who are settled therein and
must furnish temporary aid to the indigent nonresidents.
In general, a settlement is gained by residence for seven
consecutive years, during which taxes have been regularly
paid, or by payment of taxes for four years in succession
upon certain amounts of real or personal property. No
town is liable for the support of any person unless he has

gained a settlement during the ten years last preceding the date of application for relief.

The parents, grandparents, children and grandchildren of any poor person, if of sufficient ability, are liable for his support. Town paupers requiring complete maintenance, except honorably discharged veterans who must be supported outside the almshouses, as well as those without town settlement, are cared for on county poor farms, which are under the control of the county commissioners. The overseers have authority under the law to establish town almshouses, to bind out paupers, and to apprentice children. Only county almshouses are maintained. Except in certain cases, no minor between the ages of 3 and 15 years may be retained in any county almshouse.

For bringing a pauper into a town or county a penalty of a fine or imprisonment may be imposed.

In order to try to reimburse the coffers of the county poor houses, an act was approved on March 4, 1903 which said in part...

To reimburse the town or county furnishing any assistance to any person within six years preceding his death shall be entitled to recover from the estate of such person the sum or sums paid out for such assistance, the said claim to be a preferred claim against his estate after the payment of funeral charges, expense of last sickness and expenses of administration, providing he leaves no widow or minor children living at his decease.

We learn from the County Commissioners Report of 1875 that the "law of 1875 changing the settlement of paupers, has made additions to the number since Sept. 1st, as per table, all aged persons, one of them being one hundred and four years old. It is quite probable that the list will be still further increased in numbers during the year from the operation of this law." Incidentally, this 104 year old person went on to live for several more years. Her name was Betty Carr. She was a single woman from Claremont who was born Salisbury, Mass. She was 110 when she died. She is buried in the County Farm Cemetery (row 4).

The Physician's Reports begin to list the names, ages and causes of death in the reports after 1884. Here we begin to learn of all sorts of strange diseases such as marasmus, erysipelas, dropsy, consumption and old age. He also wrote in his report that year "the facilities for caring for the sick at this institution are very inadequate. The building itself is not suitable to accommodate the sick. There are no rooms set apart for that purpose. Every sick person lives in the noise and confusion, with no especial nurse but cared for by all. The building is heated by wood fires in various parts of it, so the temperature is not even. There should of course be good ventilation and a good even temperature maintained by steam heat, and there should also be facilities for bathing, and I hope that all these things will be looked after in the new building that is to be erected".

Up until this point in time, "insane paupers" were shipped to Concord to be housed in the insane asylum there at a cost to the county. It was determined that they could be cared for more economically locally therefore the County Convention of June 1883, authorized the erection of a new building for the better accommodation and care of the insane paupers and also the introduction of steam for warming the almshouse and other buildings connected therewith, all at a cost not exceeding the sum of eight thousand dollars. The reports specifies that "a building has been erected, 36x42 feet, containing eighteen rooms, besides bathrooms and closets, with all needful arrangements for the health and comfort of this unfortunate class of persons. After occupying this building for a term of nearly six months the Commissioners are confident that it is thoroughly adapted to the uses for which it was designed and constructed."

SULLIVAN COUNTY FARM BUILDINGS

This photograph shows the asylum for the insane which was built in 1885 as well as the almshouse built in 1883. The almshouse was later torn down in 1931.

The physician tells us in 1886 that "no contagious disease has occurred, nor any case of itch or other skin disease so prone to break out in places kept as homes for paupers. The addition of the new building has made it very convenient for the care of the insane and other patients needing isolation from the larger class. Those in charge have been very careful and attentive, no inmate has been neglected and every complaining one has been at once attended to. Thus no incurable disease has been permitted to reach a dangerous point unnoticed. The wards and the stairways have been kept clean and fresh, the beds and bed clothing neat, clean and in good condition, adding much to the comfort and the health of the occupants. Food appeared to be well cooked and served plentifully. There

have been but six deaths during the year ending April 31st; four males and two females, ages ranging from 62 to 78. Causes in four cases due principally to senile troubles. One had cancer and one suicide by hanging, being insane and just arrived at the farm."
FL McIntosh, MD and TE Parker, MD

The 1886 reports also discloses that the "cost of supporting the same number of insane paupers at Concord would be over $1200 more than at the farm, which shows conclusively that the erection of an insane building was a paying investment for the county."

T.E. Parker, M.D. reports in 1887 that "an unusual number of old people and those on the brink of the grave were admitted during the year, thus increasing the amount of sickness and the number of deaths; great care was given these and their sufferings mitigated as much as possible. In many of these cases complaints were raised by friends at home, claiming that poor attention was given them; whereas the fault rested on those in care beforehand, they having deprived them in a number of cases the necessaries of life. Many of these cases upon entering were pitiable looking objects and needed a large amount of patience and perseverance to make them look even semi-decent. More of such cases were added this year than in any other previous year of my attendance."

This year also saw the passing of the law that disallowed any honorably discharged veteran from being supported at an almshouse as well as their wives or dependant children. They now had to be supported in the community.

There was an apparent attempt to mitigate this tug of war between the tax payers, family and friends of the inmates and the officials in 1895 where we read in that year's report an invitation from the commissioners that reads, "We would like to have the people of the county visit the farm on any day except Sunday and see for themselves how the inmates are cared for."

According to the website of Sullivan County, "as the need to house the growing population of poor in the county increased, more land was purchased in 1869, 1903, 1907 and 1921." Apparently additional lands were needed because in the 1879 Report of the County Commissioners we find that the "last County Convention authorized the Commissioners to purchase a wood lot, for the purpose of furnishing a supply of fuel for the Alms House. Acting under this authority and the necessity of the case, a lot conveniently located, was bought in the Fall of 1878 at a cost of $450.

The County Commissioners Reports provide us a complete and accurate summary of all the financial expenditures for the Alms House throughout it's lifetime as each line item expenditure needed to be voted on. Additionally, we glean bits and pieces of the times and culture through various findings of the commissioners as well as the attending physicians throughout the years. For instance we find in the report of 1883 that "Some of the inmates, and especially the older ones, have a great aversion to water. Mrs. Wilson has informed me that it is almost impossible to get some of them to bathe. Now I recommend that it be made one of the rules of the institution that each inmate shall bathe at lease once a week in summer and once in two weeks in the winter. Personal cleanliness is conductive to good health."

In 1889 we learn that there were "only three deaths to report and all of these due to diseases incident to old age. One was insane, one deaf and idiotic and one an old hermit who was brought there in an almost dying condition." (T.E. PARKER, M.D.) The flavor of the language used has a dismissive feel to it.

Soon improvements were needed. In 1894 we learn that "The water supply has been insufficient for a long time, it having been necessary for a great deal of the time to draw water in barrels from a brook for domestic purposes. We have this past summer put in a two-inch galvanized iron pipe form the reservoir to the house, with a stand pipe to the attic. Hose has been attached to this upon each floor, and is to be kept in racks provided for immediate use in case of fire. To increase the supply of water we connected a spring that is about 90 rods above the reservoir, with the reservoir, using a three-fourth inch lead pipe. The reservoir has been full ever since this was done, and we think there is now plenty of water for all purposes. The labor of digging the ditch, about 180 rods in all, has been done entirely by the inmates and help that is regularly employed on the farm. (Residents of the poor farm were referred to as inmates. Persons in the county jail were referred to as prisoners.)

The whole expense of this improvement, including 450 feet of hose, with couplings, nozzles and racks, was $557.10. The sewers were found to be clogged up and the water setting back under the house, endangering the health of the inmates. We have had the sewers taken up and relaid, and they are now in good order. We also had an iron ventilating pipe connected and run up through the roof of the house. This was done at an expense of $122.16. The buildings on the farm are now sufficient for the requirements of the farm, with the exception of the horse barn. At some time before long, it would be well to build a new one."

The horse barn was authorized in 1896 by the County Convention. The report tells us that the new barn " is 40 by 50 feet, clapboarded and painted. The roof is slated and there is a cellar under the whole, facing the south, with a floor of concrete. It was thought best not to build on the old foundation but to build farther south in order to have the barn detached from the other buildings, on account of sanitary conditions and also on account of risk from fire. It has been thought for some time that there should be telephone connection with the alms house, as it is in such an isolated condition; in case of an emergency a telephone might be needed very much and also that is would be a great convenience at times. The Telephone Co. has constructed the line this past summer, the county paid one half the expense of construction and are to pay an annual rental hereafter. The expenses of the county farm were increased this past year by reason of losing nearly all the hogs last winter by disease."

A.L. Marden, M.D., Physician writes of the county facilities in 1895 that, "If it should be required of me to prescribe a plan of living that would ensure a person the greatest number of years of life, I would advise becoming an inmate at the County Farm. Plenty of good wholesome food is provided, regular hours for sleep, comfortable rooms, clean beds, plenty of good air and water, and careful, kind attention paid in times of illness. All these comforts, combined with perfect freedom from all the ordinary worry and anxieties that most people have in this world, are conductive to good health and longevity."

The same physician notes concern in his 1896 report that there is not a separate location for the sick, although he states that the present buildings are in "good hygienic condition".

In 1896 we learn that "the law passed two years ago obliging the Commissioners to procure boarding places for children between the ages of three and fifteen at some other place than the county almshouse has increased the expense to the county by nearly one thousand dollars a year, without, as we think, any corresponding benefit to the children. It has always been the custom in this county to get homes for the children, and it could easily be done at very small expense to the county. Children are now put in our hands that would not be under the old law, their friends knowing we are compelled to hire them boarded. In some instances, a family might be supported at the County Farm at much less expense than at their homes, but if the children must all be provided with a boarding place it is cheaper to let them remain where they are."

This problem was counteracted the following year as we learn from the Commissioners Report of 1897. "We have accomplished something this year in the way of removing people to the almshouse, who have been supported outside, at quite an expense, but there are still quite a number who ought to be taken there. We are sure, that in most cases they well be made more comfortable at the almshouse than they can be where they are now and at much less expense to the county. We ask the overseers of the poor in the several towns to aid us in our efforts to accomplish this result. We also ask the overseers of the poor to see, as far as lies in their power, that in cases where poor persons have an allowance for a certain amount of goods per week that they money is not used in an improvident manner." This was as a result of clarification of Chapter 84 of the public Statutes of New Hampshire which reads in part:

In order to correct the idea that seems to be held by many people that there is something that they call "State Aid" we print the law Chapter 85, Sections ,1,2 and 5:

Sec. 1: County paupers are those for whose support no person or town in this state is chargeable.

Sec. 2: The county commissioners may make needful orders and regulations for the removal of county paupers to the county poor farm, or any other place by them designated and no town shall be entitled to compensation for the support of a county pauper after notice of and neglect to comply with such order.

Sec. 5: The overseers of the poor shall take and transmit to the county commissioners, within 30 days after the expenditures were made, the affidavit of every pauper on whose account they were made, if the pauper is capable , otherwise of some well informed person, as to the pauper's age, place of birth, place of residence, time of residence in each place, and the time when and place where he or any of his family have been relieved or supported."

In terms of activity at the County Farm that year, the report of 1897 states that "there is nothing especially new or interesting to report from the medical department."

Some repairs were made in 1897 consisting of the laying of a hardwood floor in the dining room and shingling the Superintendent's house. Also a large hen house was built that year.

The clarification as to who should and who should not be taken to the county farm seems to have made a difference in the expenditures as we learn from the report of 1898 which states that," The expenses of supporting the poor, not at the county farm, shows a slight decrease in 1897, although by a change in the law quite a number of paupers who has been supported by the towns became county charges. The expenses of the County Farm have been more than in 1897, making an increase of the total expense for the support of the poor of $908.70. The increased expense at the County Farm is caused in part by the cost of repairs that were made. The main house was shingled, also the large barn, the wood shed and several small buildings, using more than one thousand shingles. This has all been charged in the running expenses making the cost of boarding of inmates more per week. We have had an increased number of inmates at the County Farm, but here are still quite a number that are supported at their own homes who should be taken to the farm and we again ask the overseers of the poor to assist is un the matter."

Physician's Reports are full of interesting observations and facts. For instance in 1900 we learn that there were two particularly unfortunate deaths that year. "One was an old man, who fell down stairs and injured his head so badly that he only lived a short time; the other a child who sat down in a pail of boiling water and was so badly burned that he only lived two hours." In 1901 this same physician (A.L.Marden, M.D.) stated that the "sanitary conditions in and about the buildings are as good as is possible to make them with the present arrangements. There should be water closets (toilets) and if a sufficient water supply could be obtained the cost would not be great and would be a very great improvement.

Building improvements went on as the need arose. In 1901 there was built at the farm a silo and a hen house at an expense of about four hundred dollars. "We had most of the buildings painted outside at a cost of about two hundred and fifty dollars. We also had built another silo at a small expense, so that we have all the silage that is needed for the farm. Iron ladders have been built over the alms house at an expense of one hundred and forty dollars, to take the place of wooden ladders that were taken down some time since on account of being unsafe" the account of 1902 relates.

The County Convention at the regular session in 1903 authorized the Commissioners to purchase the Stowell farm and appropriated twelve hundred dollars for the purchase. "The Commissioners have purchased the above named farm, thereby securing a supply of wood and lumber for use at the County Farm for some years, also some good land for pasturing. The Convention also authorized the Commissioners to erect a building for the storehouse at the farm. It was thought best not to build last year, "as it seemed a matter of economy, to get the lumber from the farm, and to utilize the help there as much as possible. The building will probably be erected the present year."

As promised in 1904 the report states that "a new storehouse has been erected at an expense of about two thousand dollars. The Farm contributed lumber, stone and labor for this building to amount of six hundred dollars. The size of the building is thirty by fifty feet. One half of the first floor is used as a storeroom for supplies for the house and the other half for storage for carriages and farming tools. One half of the basement has been fitted up for a cellar for vegetables and the other half for storage purposes. The entire floor of the basement is made of the best quality of cement and stone concrete."

In 1905 "by a vote of the county delegation the Commissioners were instructed and authorized to make repairs and improvements on the buildings at the county farm, as recommended by the committee to whom this matter was referred. It was found necessary to put in new foundations on one end and a part of one side of the main building. New sills and a number of new floor timbers have been put in, and some new floors. The room formerly used as a wash room has been divided into two rooms, one of which is used as a pantry which was greatly needed. A part of the foundation under the asylum has been laid over and some new floor timbers put in. Extensive repairs and additions have been made on the system of steam heating and to the water fixtures throughout the buildings. A steam laundry has been installed in the basement on the west wing, formerly used for the insane. The machinery consists of a twelve horse power boiler, a six horse power engine, a one hundred and fifty shirt washer, extractor, dryer with the necessary set tubs, shafting, etc. It has proved to be a great labor saving investment. A great improvement has been made in the sanitary arrangements of the institution. A toilet room with a lavatory and bath tub has been put in the superintendent's cottage, a toilet room has been installed on each floor of both the men's and women's wards, with a bath tub in each room on first floor. At the asylum a toilet room with a bath tub has been put into each ward on first floor. It is an improvement which not only adds to the comfort of the inmates but to their healthfulness as well."

With each improvement comes additional expenditures to support the new arrangements. For instance in 1906 it was determined that "the present system has proved to be inadequate the past season to supply the water required for the new steam laundry, bath and toilet rooms installed during the preceding year. It is very important, also, that a supply be secured for use in case of a fire, and that hydrants be provided at suitable places outside of the buildings. The almshouse is equipped with standpipes, but no supply of water which would last any length of time, is now available. The rare of insurance on the farm buildings is $31.00 per thousand yearly, and it is hoped a substantial reduction can be secured when a suitable water supply is provided. Surveys and estimates have been made for the expense of building a reservoir, laying pipes and placing hydrants where needed, and the county convention will be asked to consider the whole matter then they meet at the farm."

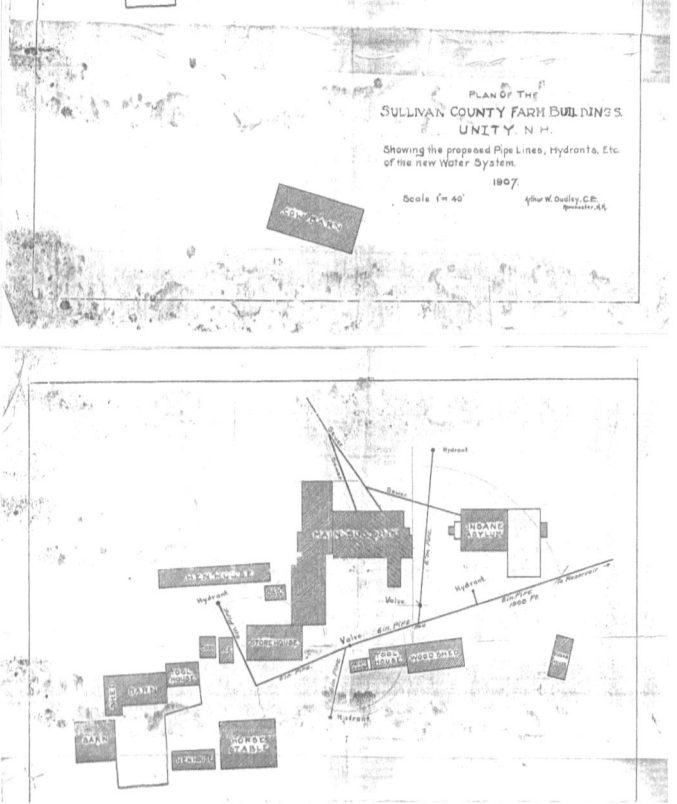

The jail property located at the County Farm also required updating in 1907. The jailer's residence was wired for electric lights and steam heat was put in. The roof on both the house and barn were slated. The buildings were painted and new flooring laid.

JAIL AND RESIDENCE.

Additionally, new wood floors were laid in five rooms of the almshouse and a new large contraption called a refrigerator was built in the basement to hold 1000 pounds or more of meat.

This same year also saw the dreams of the reservoir come to fruition. "The water system for which an appropriation was made by the County Convention has been put in this season. Plans and specifications were made by Arthur W. Dudley, C.E., of Manchester, who also had charge of the construction of the system. Bids were solicited for the building of the dam, excavating, laying pipe, setting of hydrants, etc., the pipe and hydrants being furnished by the Commissioners. The contract was awarded to Ablett & Bowes of Cohoes, N.Y., who fulfilled their contract to the satisfaction of the Commissioners and Engineer. A dam 80 feet in length, 14 feet high, using over 200 barrels of cement, was built across the brook not far from the old reservoir, flowing a pond holding about 500,000 gallons. The elevation is nearly 100 feet higher than the buildings and gives a pressure at the Almshouse of about 45 pounds. One thousand nine hundred and twenty-five feet of 8 inch and 600 feet of 6 inch pipe were laid connecting with four hydrants so located as to give good fire protection to all buildings at the farm. The bath rooms, boiler, laundry and barns are also connected, with the new system, the old system being kept for domestic use only. Before the work was accepted by the Commissioners, Chief Engineer Sears of the Claremont Fire Department, made a test of the system and pronounced it a success water being easily thrown over any building on the institution. Six hundred feet of 2 ½ inch and 300 feet of 1 ½ inch hose for protection against fire have been purchased. The farm now has the best water system of any institution of its kind in the state. The rate of insurance has been reduced 6 10ths of 1 per cent. The reservoir will furnish the institution with ice which will save expense, as the supply formerly had to be procured several miles from the farm."

RESERVOIR FOR WATER SUPPLY, CONSTRUCTED IN 1907.

The year also saw the installation of a "marble marker at each unmarked grave of those who have died at the Farm, and been buried in the County cemetery within the past few years to the number of eighteen. (No sign of these markers remains however.)

Modern fire escapes were installed in 1908 and the physician reported in 1909 that "Too much cannot be said in favor of the sanitary conditions as they now prevail at the farm; the water supply is pure and abundant; the buildings are connected with an adequate sewer; the floors, household furnishings and clothing are kept scrupulously clean; and this with well cooked and wholesome good, which is always supplied in abundance, helps to bring about a general condition of good health."

With the water in good order now attention was turned toward the condition of the electric lighting in 1909. "Consequently an agreement was made with the Claremont Power Company whereby the county owns the line from Claremont village to the farm, and the company furnishes the current for lights and power at regular meter rates. The system consists of 120 inside lights conveniently located throughout the cottage, almshouse, hospital, laundry, barns and storehouses and five 32 candle power lights outside the buildings. This arrangement has proved very convenient and satisfactory."

Still there was more to do. In 1910 we learn that "during the past season a new stock barn has been built at the farm 40 x 84 feet, with basement under the whole, with cement walls for foundations and slate roof. The main floor contains stables on either side with box stalls and granary. The stables are equipped with 42 of the "Perfection Swing Stanchions" which is a great improvement over the former was of confining cattle. One of the old barns has been removed to a suitable place west of the horse barn and repaired and remodeled into a tool house and slaughter room, the later being fitted up with modern conveniences for butchering. While improvements have constantly been going on about the County buildings for the past six years, there yet are some things still needed to bring the same to a reasonable state of perfection at the farm, among which are the remodeling of the Superintendent's house, and the building of a new heating plant separate from the main buildings ,as recommended by the last report of the State Board of Charities and Corrections."

In 1911 the Commissioners reminded the citizens that a new Superintendent's house would be needed. Finally in 1913 by a vote of the County Convention, the County Commissioners were authorized to construct a new house for the use of the Superintendent and employees at the County Farm. Hira R. Beckwith, architect, at Claremont, was employed to make plans and specifications and to have general charge of the work. After the plans and specifications were prepared the Commissioners advertised for bids for carpenter and mason work, above the foundation, all the material to be furnished by the County. All bids submitted, were in the opinion of the architect and Commissioners, too high, and it was decided to hire help by the day to build the house.

The amount paid for carpenter and mason work was about three hundred dollars less than the lowest bid for such work.

The cottage which heretofore had been used for the home of the Superintendent and employees was moved onto the lawn, and used as a home, while the new house was being built.

The new house stands on the same site as the old one did, excepting, that it extends eight feet further out on the west side making a building thirty-six feet in width, seventy-two feet in length, with flat roof, covered with tar and gravel. A piazza extends across the north end and for about thirty-six feet on the east side of the building.

The house has on the first floor a living room, sleeping room, dining room, office, kitchen pantry, workroom, storeroom, milkroom and toilet. On second floor six sleeping rooms with a closet for each, ample hallways and a toilet room.

The finish throughout the house is of hard or yellow pine, and the floors are all of hard wood.

The house is heated by steam, lighted with electricity and has modern plumbing throughout. The entire construction is of good material and workmanship.

NEW SUPERINTENDENT'S HOUSE, COUNTY FARM

Views of the Administration Building and Surrounding Grounds

With the new Superintendent's home completed it was time to turn attention to the heating system at the Almshouse. The 1914 report states, "It will not be very long before some considerable outlay will be necessary for a new heating plant at the Almshouse. The system now in use was installed in 1885. The boiler is under the east end of the building occupied by the inmates as living and sleeping rooms. The Secretary of State Board of Charities has recommended a change for some years, and in the last annual report of the Board the matter was referred to and the statement closes by saying the present conditions, make the danger to life from fire great." The necessity for change came in May of 1915 when "the boiler inspector condemned the heating boiler." This was an enormous and expensive undertaking. The 1915 Commissioners report states that " a special meeting of the County Convention was called for June 22 and the convention voted to authorize the County Commissioners to borrow money on the credit of the County, the sum not to exceed Fourteen Thousand Dollars, for repairing the Almshouse, to build a woodshed, boiler house and chimney and to install suitable boilers, pipe lines and radiators for heating all the buildings where heat is required. The boiler house was built on the West side of the Superintendent's house, 30 feet distant. The building is forty-three feet long by twenty-five feet in width with basement, twelve feet in depth, with concrete walls for sides to the surface of the ground and ten feet of brick walls above. The roof is of tar and gravel construction. F.E. Newcomb of Newport furnished the boilers, piping, radiators and all the fixtures for a complete heating system and did the work of installing the same. The system seems to be amply equal for heating all the buildings and has worked with entire satisfaction. The sum expended for the improvements authorized by the convention was Twelve Thousand Dollars."

Lumber, wood and water needs continues to plague the Commissioners. In 1918 we learn that "the County Delegation was called together on March 14, 1918, to consider the question of purchasing the Charles Willard lot, containing about two hundred acres of land. This lot is located about one mile from the County farm and the object is purchasing this land was on account of a large lot of wood and timber standing thereon and valuable springs of water." This lot was purchased on credit for $4250.00 at an interest rate of 5%.

Socially, things were changing. World War I was over and prohibition was enacted. In 1919 the Jailer's Report states that, "The restrictions placed by law upon alcoholic beverages have done more than any single cause to reduce the criminality in this County. The same experience elsewhere tends to show that the organized traffic in alcoholic beverages leads directly to crime and is one of the most formidable menaces that society has had to contend with." Prohibition was officially begun on January 16, 1920. Sullivan County was proactive in this regard. On May 3, 1916 Frederick Aiken, County Commissioner filed this report.

"On January 10, 1916 I went to the County Jail, taking with me the order of the court on which this return is made and received from E.J. Deming, Deputy Sheriff and Jailor, the following list of Malt and Spirituous liquors, also certain empty bottles, jugs, casks as designated below:

57 bottles of malt liquor known as Ppaff's Beer
21 " " " " " " " Norfolk Brewing extra Golden Ale
12 " " " " " " " Budweiser Beer
¼ keg " " " " " Lager Beer
8 bottles of spirituous liquor known as Buckeye Pineapple Horehound Rock & Rye
1 bottle of spirituous liquor known as Whiskey
2 ¼ bottles of Martini Cocktail
1 bottle of DuBury Cognac
1/16 of a bottle of Manhattan Cocktail
1/3 bottle of alcohol

1 empty bottle listed as Sloe Gin, large size, no label
1 empty bottle listed as Garden Gin, dry
1 empty bottle listed as Londonderry Gin, dry
1 empty bottle listed as Black and White Scotch Whiskey
1 empty bottle listed as Sloe Gin
1 empty gallon jug listed as wine
1 empty gallon jug listed as Rum
1 empty gallon jug listed as Fair View Whiskey
1 empty cask of Hoffman's House Whiskey
1 empty cask of Kays Tom Gin
1 empty gallon jug of spirituous liquor
1 empty unlabelled bottle listed as wine
1 empty bottle listed as Old Rum Fruit Punch

1 empty bottle listed as Port Wine
1 empty bottle listed as Giralda Brandy
1 empty bottle listed as Ginger

Times were quiet at the farm for a couple of years. The only news in 1921 was that the crops were good with the exception of the hay and a new cement floor was added in the basement of the new cow barn. However, after reviewing the financial needs for the next two years it was decided that it would be advantageous to purchase the Judkins farm situated about one and one half miles West which contained a "house, granary, two barns, about 300 acres of land and an estimated 1,000 cords of wood and quite a quantity of oak and chestnut lumber." The Commissioners purchased the farm for $2100.00.

Each year gave a financial accounting of where moneys came from and how they were spent. In 1923 we learn that the Red Men paid $42.00 for the board of Herbert Way who is buried in the County Farm cemetery. "The Improved Order of Red Men traces its origin from the Sons of Liberty patriots who were active before the American Revolutionary War and are well known for their participation in the Boston Tea Party. The Sons of Liberty and in turn the Improved order of the Red Men emulated in their organization, philosophy and regalia the League of the Iroquois or "Six Nations". The crowning feature of the League as a political structure, was the perfect independent individuality of the national sovereignties, in the midst of a central and embracing government adequate to deal with the internal affairs and powerful enough to conquer all other Indian nations with which it came in contact. Hospitality was extremely important. The Iroquois would surrender his dinner to feed the hungry, vacate his bed to refresh the weary and give up his apparel to clothe the naked. Their eventual decline, however, was due primarily to the introduction of "fire water" through the European traders." A Nation of Red Men by Kathleen O'Connor). In the New Hampshire IORM Improved Order of the Red Men State annual meeting proceedings for 1924, Herbert G. Way is on the Fraternal dead list for Skitchawaug Tribe No. 29. It does show him as dying in September 1923. His is listed as a past Sachem (president) of the tribe at one time.

More building headaches were soon to be on the horizon. "The duties at the Institution have been greater than usual by reason of the fact that during the fall of 1929 our Almshouse at Unity was visited by the State Authorities who informed us that the same was not in proper condition for the best welfare and safety of the inmates. After carefully considering the matter, the Commissioners called the County Delegation together. A thorough investigation of existing conditions was made by all members of the Delegation, and it was decided by the Convention that it would not be wise to remodel the present buildings." The 1930 Commissioner Report contained the Report of the building Committee which reads, "The Sullivan County Delegation at a meeting held April 12, 1930, voted to build a new Almshouse in Unity, as recommended by the investigating committee, in the preliminary plans which were submitted, and a bond issue of $150,000 was authorized for the construction of the building and its furnishings. The Committee elected to have charge of the work employed Wells & Hudson of Hanover, as architects, and awarded the contract for construction to Davidson & Swanburg of Manchester. The new fire resisting building situated south-east of the old almshouse, is constructed of brick, cement and steel, with inside partitions of terra cotta tile and gypsum tile, and is at the present time nearing completion. It is 192 ft. by 56 ft., two stories above ground and a basement which is fully utilized. In the basement are located the kitchen, dining rooms, laundry and other equipment necessary for the institution. In one end of the basement, and entirely separated by solid walls, from the other rooms, is located the House of Correction. An underground cement tunnel connects the basement with the Superintendent's home and also conveys the steam pipes from the heating plant to the new building.

On the upper floors, besides the inmates rooms, are the office and reception room, chapel, hospital department, rooms for nurses, a well equipped infirmary and diet kitchen".

NEW BUILDING—SULLIVAN COUNTY HOME

The following year "benefits" for staff were initiated. In 1931 we learn that to express gratitude for services rendered from Superintendent and Mrs. Grout an annual vacation would be awarded them without loss of salary. This was quite a vote of confidence since the depression was upon the country. There was grumbling that year about the budget and the Commissioners addressed this in their report with the comment, "Everyone knows that the unemployment situation at present is the worst it has ever been but people need to live like human beings." Constant battles between the Overseers of the poor and the Commissioners resulted that year in Paul A. Mineau from Claremont resigning his position.

Also that year Article III Report of Committees suggested a new name was in order for the Sullivan County Farm. "Mrs. Hamlin of Charlestown suggested the following new names: (1) Hillcrest (2) Mountain View (3) Mountain View Home (4) Sullivan County Home. Sullivan County Home was adopted as the new name and would replace the old. Incidental expenses for this very momentous year included moving the old jail house and making improvements, increasing the capacity of the reservoir, removal of the old almshouse, repairs of the Superintendent's building, grading, etc. at a cost of $7813.25." The new road constructed from the Claremont town line to the Sullivan County Home was now complete.

The old jail building would become the new children's house. The 1932 report explains: "Our slaughterhouse and pens for brood sows were not over seventy-five feet from the Children's Building and without explanation one should know the result. We have under construction a slaughter house thirty feet by thirty feet, with brood pens, thirty by one hundred feet attached. This space will provide twenty pens, which will be a great help, as we have an exceptional market for young pigs, with a growing demand. We have also built one small brooder house, making four in all. Last summer all the buildings were painted, the only expense being the cost of the material, the town workers doing the work. With the labor being furnished by the town workers and under the supervision of our carpenter, all this work had been done with very little expenditure of money, as the material, with the exception of nails, building paper, cement, and the small items of hardware, were practically all furnished from the Farm. These improvements have seemed a necessity and were accomplished with as little actual money expense as possible. We have about six hundred cords of four foot wood on hand. This wood and about 90 thousand feet of logs were cut during the winter. We have kept two or three gangs of men in the pastures cutting brush and in the pine woods trimming out and cutting lower dead limbs, greatly improving the pine lots. Considerable work has been done in the interior of the buildings, renovating furniture and painting. Because of the increase in the number of hospital patients it was thought necessary to move the sewing room to the first floor of the new building, thus giving us a large room for patients. The hospital diet-kitchen was in the front of the building, some distance from the food elevator. This room was also too small to accommodate the increased number of trays. To remedy these deficiencies, a larger room, more accessible to the food-elevator, was converted into a diet-kitchen. In order to qualify as a Lying-In Hospital, the State required us to have a nursery; also the number of babies has been increasing and it was no longer possible to keep them in the operation-room, as sometimes we have had four or five at one time. Therefore, the former diet-kitchen was made over into a very adequate nursery. The Children's building, partially completed during 1931, was finished during the first of this year, with perhaps some expense but we feel it was money well spent, as now we have quarters for children, surpassed by few. The building was formerly the jail and was completely remodeled, most of the material being used being salvaged from the old building. We can now accommodate about 35 children, having separate dormitories for girls and boys and also separate play grounds, with swings, slides, etc. We have tried to carry on with as little expense as possible to the County, and the people must realize that with the increase in admissions to the home, feeding an average of about forty workers from Claremont daily, and with the average of over thirty patients in the hospital, there is bound to be a marked advance in costs over the previous years."

Also in 1932 we learn that "the greater part of the spring was devoted to grading and beautifying the grounds, setting out trees, shrubbery, etc. Our prisoners were engaged much of the summer in constructing roads around the buildings, using stone walls as a base, covered with gravel. Altogether they built more than three-quarters of a mile of road, thereby not only making the roads better but also disposing of a large number of rocks and unsightly stone walls. We have a large number of workers coming from Claremont and they are taken to and from the Farm in our trucks. These men have been a great help for we have accomplished much that would not have been possible with our regular help. Besides building roads, grading, etc. we have erected at a very low money cost, several buildings that were badly needed. We found it imperative to remove our carpenter and blacksmith shop from its location under the Children's Building, as the smoke from the forge was damaging the rooms above and the noise from the machinery was very annoying. Consequently we erected a building forty-two feet square to accommodate both the shops; a building well constructed, with a lumber loft above. This has proved a great help in our work and was built at a very low money cost. A place for storing vegetables was also much needed as the room in the cellar, the only place available, was too small and the steam-pipes running through it made the room so warm that the vegetables spoiled. Therefore, we constructed a vegetable cellar, fifty by thirty feet, built of stone and cement.

Recipients of county aid, most of whom are from Claremont have done a large part of the work. They have cleared away and burned the brush in the pastures leaving a larger area for feed for the cattle. No effort has been spared to bring the land into a highly productive state."

"Under the auspices of the N.H. Unemployment Relief Committee two trained investigators came into the county to help adjust the case load made heavier than ever before because of the prolonged unemployment. Miss Margaret Price was stationed in the Overseer of the Poor's office in Claremont in April 1932 and Mrs. Edna Whitney was placed in the County Commissioner's office in Newport in June, where she remained until December when Miss Mildred Israel substituted. They have done good work. There is now a complete record of county aided case on file in the Claremont and Newport offices."

The Commissioner's Report of 1933 tells us, "Several minor improvements have been made at the Home. The only major project of the year was the start of a new water system. The shortage of water in August 1932 and again in June and July 1933, made it necessary to take advantage of the vote of the Delegation appropriation of $15,000, upon plans and recommendations submitted to the Delegation at the February meeting held at the County Home, by Mr. Loyal Barton, Civil Engineer, the plans and estimated being made by Mr. Barton at the request of the former board of Commissioners. Work was started July 26th, using all the county dependents for the common labor. All material was purchased in the County at the lowest bid. Every difficulty known to man was encountered during the progress of the work. Early in September it was made evident that the appropriation would not complete the project and a special meeting of the Delegation was called for September 26th. After an inspection of the Water Project by the Delegation, the following vote was passed: That the sum of three thousand dollars be appropriated to carry the water project to a temporary completion, with the understanding that the project will be completed in 1934. This added appropriation carried the project until the cold weather. On November 20, the Commissioners received a contribution from C.W.A. Government Funds, to employ 20 men, a timekeeper and foreman, and to purchase certain materials. With this assistance, the project was continued until February 15, 1934 using Claremont dependents for the labor. The County had to furnish certain labor, materials and trucking in order to receive the C.W.A. assistance, amounting to approximately $1300.00. It is the hope of the Commissioners that more assistance from C.W.A. funds will be available in order that the whole project may be completed by May 15. The County will be asked to furnish all materials under the next C.W.A. allotment.

The Civil Works Administration (CWA) was established by Roosevelt's New Deal (beginning 11-8-1933) during the Great Depression to create jobs for millions of unemployed. The jobs were construction mainly improving or constructing buildings and bridges. It ended on March 31, 1934 after costing $200 million a month.

On July 1, 1933, the State took over the supervision of all outside relief cases under what was known during the last session of the legislature as house Bill No. 417 and in now known as Chapter 160, Laws of 1933. This law set up a Director of Poor Relief appointed by the governor, who has general charge of all outside poor relief in the State. The Director divided the State into districts, but not cutting across county boundaries and appointed a trained welfare worker known as the County or District Supervisor. The Supervisor had general charge of the administration of all outside poor relief in that county or district. "All employed, excepting the supervisors, shall be residents of New Hampshire. Under Section 4 of the law, supervisors are directed to consult with and employ in the administration of the act, County Commissioners, Selectmen, Overseers of the Poor, and other agencies, so long as reasonably competent services are rendered by the above officers. Under the act, the Governor and Council are authorized to receive from the Federal Government or other agencies, any money advanced to, or placed at the disposal of the State for the relief of the distressed. Also the Governor and Council are authorized for the purpose of financing this act, to issue bonds to the amount of $600,000 for the balance of 1933 and $1,200,000 for the year 1934. All moneys so received, are directed under section 7 to be deposited with the State Treasurer and are to be known as the Emergency Relief Fund and to be paid out only on warrants drawn by the Governor for purposes of this act alone. All expenditures for outside relief are, in the first instance, paid out of the Emergency Relief Fund, and then from time to time, the cities, towns and counties reimburse the State to the extent of 25 per cent of the total cost of outside poor relief."

Outside Poor Relief continues to be supervised by the State during 1934. The money was supplied as follows: Government and State approximately $72,000, Sullivan County $24,000. "Several minor and two major improvements have been made at the Home. The first major improvement was the completion of the Water Project which was started in 1933. The payroll for the labor was furnished by C.W.A. and F.E.R.A. entirely, the County supplying the money for supervision, engineering, materials and trucking. The total cost to the County for the year 1934 was $4,079.73. This includes the water-proofing of the reservoir which was not contemplated at the time of the Delegation meeting, March 1, 1934. The project was completed in October and has now been in operation three months and is most satisfactory in every way. The second major project was the installation of the sprinkler system in the new building and addition to the hydrant service around the buildings. This was completed in November at a cost to the County of $3,100 for the sprinkler system and $750 for the hydrant extension. At the completion of these two major projects, the insurance on all the buildings at the County Home was completely readjusted, some insurance was added on the old buildings, at a yearly savings of approximately $600 in premiums, and at the same time giving additional fire protection."

The Federal Emergency Relief Administration was the name given by the Roosevelt Administration to a program whose goal was to eliminate adult unemployment. From when it began in May 1933 until it closed its operations in December 1935, it gave states and localities $3.1 billion to operate local work projects and transient programs.

In 1935 "it was voted to approve the action of the County Commissioners, in purchasing a wood lot in the town of Unity, for the sum of $600. The same being bought in 1934." Also that year, the Social Security Act created a program called Old Age Assistance, which gave case payments to poor elderly people. This one Act effectively eliminated the almshouse system over the next few years.

Building went on in 1936 the Commissioners tell us, "We have just completed a two-story hen-house, seventy-five by twenty feet, with running water and large enough to accommodate about seven hundred hens. This was done because one of the hen-houses was old and much in need of repair and was located too near the Annex, a building housing some of the charges. The tool and wagon shed has been enlarged by about fifty feet. An extension about fifty by thirty feet has been added to the piggery."

Continuing bad economic times saw an increase in the County load of dependents at a rate of 59% between November 1, 1937 to January 1, 1938. The 1937 Commissioners Report of a meeting that was held on

March 22 at the County Home, "the morning was spent in inspecting the barns and stock. The sows and little pigs appealed to Brigham (one of the Commissioners) and one of the barred rocks laid an egg in his pocket." The report of 1938 shows no improvement in the economic conditions of things. Again it reports an increase the County load and states that two major improvements should occur, the building of a fireproof garage and a new jail, "if and when the County is in a financial position to make them." It was moved that the County delegation ask the County Commissioners to cooperate with the Town selectmen in putting County relief cases on Town WPA projects and the commissioners be authorized to make reasonable compensation to the Towns to take care of the increased cost. The vote was unanimous.

The hurricane on 1938 came causing more trouble for the County home. The report of that year states, "The expenditures have been perhaps somewhat high but in checking the extra work necessitated by the hurricane we consider ourselves fortunate in getting by as well as we have. Starting last spring we transplanted about fifty thousand spruce and pine seedlings on land that had been cleared for that purpose. During the summer, the cemetery was entirely graded and seeded, the stone walls re-laid, and new gates built. Many of the stones on the unity side of the cemetery had fallen and these were re-set. This represents a great deal of work but the result was worth all that was put into it.

Practically all the buildings were painted during the summer and fall, necessitating considerable expense. During the winter we have thoroughly renovated the Annex, refinishing all the floors. The main building and the Superintendent's residence have also been painted. More or less of this inside work is done every winter. We have had a crew of W.P.A. workers who have helped greatly in cleaning up after the hurricane. We have drawn to our mill and sawed about two hundred thousand feet of logs and no doubt we have at this time as much, if not more still in the woods. We had miles of fences to repair after the hurricane. This required a large amount of barbed wire and thousands of fence posts. A large wagon storage shed and a bridge leading to the cow barn were completely wrecked and had to be re-built, both requiring new metal roofing, hardware and several tons of cement. The metal roof was blown from the saw-mill and had to be replaced. The roof of the main building was also badly damaged and considerable slate blown rorm other buildings. Two chimneys were blown down, breaking through the roof. In a part of the old piggery, a floor which was badly rotted had to be torn out and replaced with a cement floor fifty by thirty feet. All cement floors in the piggery have been covered with boards, as we found that hogs were becoming crippled to a certain extent by standing on cement. We have started the construction of a ramp or walk in the rear of the main building. This walk will be connected with the driveway and extend to the hospital on the second floor, permitting us to convey patients to hospital by wheel-chair or stretcher. This will be a great help, as we now have to carry patients up two flights of stairs. We have purchased a stretcher for transportation of patients, and this, with the specialty constructed interior of our new Chevrolet suburban, which can be converted into an ambulance, is a great convenience. It furnishes a comfortable means of transportation for patients and has already more than paid for itself. A cement mixer was purchased during the year. This has been much needed and we find it a great help in construction work. There will be quite a change in our Board of Commissioners during the coming year and we are anticipating the same friendly cooperation we have always received from the present Board. We of course expect new views and suggestions, which are always welcome, and we sincerely hope we may be able to carry out any ideas with satisfaction to the Board. To the retiring members of the Board we wish to express our appreciation of their interest and help, and to them we extend our best wishes."

Identification of those in the Sullivan County Almshouse Cemetery

Row One (from front to back, left to right)

Herman Jackson 7-10-1913

Helen May Mason 1926-1927
Died 10/4/1927
Born 1/13/1925
Cause of death: pneumonia
Admitted November 26, 1926
Born in Cornish
Sending town: Cornish
Age: 1

Lillian Kimball 1888-1929
date of death:1/29/1929
date of birth: not listed
cause of death: pneumonia
Admitted 1-13-1926 (along with Amasa Kimball age 61 also of Cornish)
born in: Canada
sending town: Cornish
age at death: 41

William Hill 1851-1929

William Johnson
1853-1929

Alexander Monroe 1875-1929

Mary Pratt 1865-1929
Died:10/20/1929
Born in: Alstead
Sending town: Charlestown
Died of: arteriosclerosis
Age: 63
Admitted 10/26/1916

Arthur Page died 1929
Died:10/23/1929
Born in :Sunapee
Sending town: Sunapee
Died of: arteriosclerosis
Age:69
Listed as a boarder

Mary Mackie died 1930
Died:5/28/1930
Born in: Finland
Sending Town: Newport
Died of: chronic arthritis
Age at death: about 60
Date of admission 4/24/1928

Carl McKeeser
Date of death: 10/23/1930
Born in: New Bedford
Sending Town: Claremont
Died of: pneumonia
Age: 84
Admitted 6/27/1929
d.o.b.:4-12-1846

Florence Kimball 1867-1931
Admitted: 2/3/1913
Age: 46
Born in: Canada
Sending Town: Plainfield
Widow

Carroll Scott died 1932
Died: 3/10/1932
Born in: Croydon
Sending Town: Croydon
Died of:chronic myocarditis
Age:79
In the record his name is listed as Scott Carroll and he was not "discharged" until 3-18-1932

George Sanderson 1839-1932
Died:6/28/1932
Born in: Woodstock, Vermont
Sending Town: Plainfield
Died of: angina pectoris
Age at death: 53
Admitted 4/15/1932

Edward Wilson 1887-1932
Date of Death: 10/30/1932
Born in: Norfolk, Virginia
Sending Town: Claremont
Died of: cerebral hemorrhage
Age:55
Admitted 5/9/1931

Fred Bashaw
Admitted: 1/17/1895
Sending Town: Charlestown
Born in: Charlestown
Cause of Death: chronic nephritis
Age: 52

Baby Joan Quimby
Baby Joseph Quimby
Newborn 1933
Date of death: 9/2/1933
Date of Birth: 9/2/1933
Born in:Unity
Sending Town: Unity
Cause of Death: premature birth
Joan lived 3 ½ hours, Joseph was stillborn; mother's name
Blanche Quimby age 31 from Newport born in Canada
admitted 9/2/1933 discharged 9/16/1933

George Mercier
1880-1933
Date of Death: 9/4/1933
Born in: Canada
Sending Town: Sunapee
Cause of Death: chronic myocarditis
Age:53
Admitted 6/13/1933

John O'Brien
Date of Death: 12/28/1933
Born in: Langdon
Sending Town: Langdon
Cause of death: general arteriosclerosis
Age at death:73
1860-1933
Listed as a boarder

Douglas McInnis
1881-1934
73
Date of Death: 2/2/1934
Born in: Nova Scotia
Sending Town: Sunapee
Cause of death: chronic myocarditis

Norman Robert Wright
1934-1934
Date of Death: 3/30/1934
Born in: Claremont
Sending town: Claremont
Cause of Death: syphilis
Age at death: 7 weeks
Admitted March 26, 1934

Charles Williams
1846-1934
Date of Death: 12/2/1934
Born in: Buckland, Mass
Sending Town: Sunapee
Cause of Death: bronchopneumonia
Age: 88
Admitted 11/30/1934 as a boarder

Raymond Dutilly
1935
no information given in the Commissioner's reports for
1934 or 1935

Howe Baby
1935
Date of Death: 11/17/1935
Date of birth: 11/4/1935
Born at: Sullivan County Home
Age: 2 weeks
Sylvia Mae Howe, mother Isabel A. Howe, age 24, born
in Newport but from Goshen, admitted 6/6/1935

Eugene Cobleigh
1854-1936
Date of Death: 3/16/1936
Born in: Gardner, Mass.
Sending Town: Plainfield
Age: 81
Admitted 5/9/1935

Amasa Kimball
1863-1936
Date of Death: 10/13/1936
Born in: Plainfield
Sending Town: Cornish
Age at Death: 68
Admitted 1/13/1926

Elbridge Amsden
1-1937
Admitted 12/5/1921
Date of Death: 4/6/1937
Born in: Claremont
Sending Town: Claremont
Age: 91

Sobolvesky Baby
1937
No information is listed about this baby but we do know
that the mother's name was Dora Sobolvesky who was 37.
She was admitted from Newport on 5/17/1937 and was
discharged on 5/26/1937. Therefore, it is reasonable to
assume that the child was born during that time period.

Michael Gobin
1860-1937
Admitted 7/7/1937
Date of Death: 11-19-1937
Sending Town: Newport
Age 77

Shirley Stearnes
1938
Date of Death: 6/12/1938
Date of Birth: 6/10/1938
Born in: Grantham
Age at death: 3 days
Admitted at 1 day old

Ernestine M. Miller
1939
No information is listed in the Commissioner's Reports
about anyone of this name

Edward Southworth
1850-1939
Date of Death: 10/12/1939
Sending Town: Unity
Age: 89
Admitted 6/1/1938

Smith Baby
1939

No information provided either in the ledger or the
Commissioner's Report

Everett Colburn
1884-1940

Row Two

Samuel Reynolds
3/25/1912
54
Admitted: 12-16-1911
Born In: Canada
Sending Town: Claremont
Cause of death: heart disease
Listed in the reports as Samuel Renolds

Josiah Davis
12/6/1901
Admitted: 6-3-1900
Born in: Sunapee
Sending Town: Sunapee
Married
Cause of Death: Cancer
Age: 75

Hannah Joyce
8/26/1903
76
Admitted: 12-22-1897
Born in: Croydon
Sending Town: Croydon
Was not discharged until 8-28-1903
Cause of Death: disease of the liver

Adam Brooks
1/02/1903
25
Admitted: 10-21-1902
Age: 25
Born in: Putney, Vermont
Sending Town: Claremont
Married
Cause of Death: Bright's disease

Richard Heywood
7/4/1903
74
Admitted: 12-9-1894
Born in: Charlestown
Sending Town: Charlestown
Cause of Death: disease of the heart
Single

George Roberts
1/2/1903
33
Admitted: 7-23-1902
Born in: Sunapee
Sending Town: Newport
Cause of Death: Cancer

John Hosmer
8/26/1903
62
Admitted: 2-8-1900
Born in: Langdon
Sending Town: Windsor
Cause of Death: Heart disease
Single

Abbie Chase
1/22/1904
55
Admitted March 7, 1868
Born in: Croydon
Sending Town: Croydon
Cause of Death: eosysipelas

Cordelia Warren
6/11/1904
83
Born in: Hampton, Vermont
Sending Town: Newport
Single
Admitted May 5, 1899

Peter Thompson
8/2/1904
83
Born in: Scotland
Sending Town: Claremont
Single
Cause of Death: cholera morbius
Admitted September 3, 1901

Mary Hammond
2/14/1905
56
Admitted: 10-4-1880 at age 31
Born in: Grantham
Sending Town: Grantham
Single
Cause of Death: apoplexy

Livera Chase
7/23/1905
82
Admitted: 10-17-1870 at age 47
Born in: Croydon
Sending Town: Croydon
Widow
Cause of Death: apoplexy and old age

Charles Jebb
9/3/1905
78
Admitted: 12-7-1903 at age 76
Born in: England
Sending Town: Newport
Cause of Death: heart failure and old age
Married

Sarah Sleeper
1/20/1910
67
Admitted July 22, 1902, age 59
Born in: Hardwick, Vermont
Sending Town: Springfield
Cause of Death: pneumonia

Josiah Sleeper
5/14/1909
70
Admitted: 7-22-1902, age 63
Born in: Canaan
Sending Town: Springfield
Cause of Death: strangulation due to food inhalation

John Snow
7/30/1910
62
Born in: Bedford, New Hampshire
Sending Town: Washington
Cause of Death: heart disease
Admitted 3/19/1910

William Thompson
9/9/1910
72
Admitted 1/20/1908
Born in: England
Sending Town: Claremont
Cause of Death: heart disease

Edward Hallihan
9/14/1910
84
Admitted 11/7/1871
Born in: Ireland
Sending Town: Langdon
Cause of Death: arteriosclerosis

John Dugan
11/26/1910
53
Admitted: 1/14/1910
Born in: Ireland
Sending Town: Claremont
Cause of Death: hung self
Single

Hamilton Miller
4/12/1911
47
Admitted: 3-16-1900
Born in: Ireland
Sending Town: Claremont
Cause of Death: congestion of the lungs and throat

John Hurley
7/7/1912
90
Admitted 5-22-1911
Born in: Portland, Maine
Sending Town: Springfield
Cause of Death: senile dementia
Name is spelled Hurleigh in the report

Sylvanus Olney
12/6/1913
90
Admitted Aug. 1, 1911
Born in: Canada
Sending Town: Cornish
Cause of Death: senility

Mildred Davis
11/2/1915
11
Born in: Windsor, Vermont
Sending Town: Plainfield
Was admitted 10-31-1914 and was discharged 4-24-1915;
"Went to the Feeble Minded Home in Laconia

Michael Sullivan
4/19/1916
53
Admitted: 2-16-1916
Born in: England
Sending Town: Claremont
Cause of Death: valvular of the heart

Mary Estabrook
6/23/1917
73
Admitted: 4-11-1899 at age 55
Born in: Claremont
Sending Town: Newport
Cause of Death: duodenum

Fred McDonald
12/23/1917
63
Admitted: 7-3-1915 at age 61
Born in: Nova Scotia
Sending Town: Newport
Cause of Death: arteriosclerosis

John Hubbell
5/21/1918
78
Admitted: 12-9-1884 at age 44
Born in: Newport
Sending Town: Newport
Cause of Death: cerebral hemorrhage

George Lewis
6/19/1919
53
Date of admission 5/27/1919
Born in: Massachusetts
Sending Town: Newport
Cause of Death: tuberculosis

William Miller
7/14/1920
Age 72
Was a boarder who entered 7/1/1920
Born in: Scotland
Sending Town: Claremont
Cause of Death: Bright's disease

Frank Labarge
5/22/1921
31
Admitted: 9-2-1920
Married
Born in: Peterborough
Sending Town: Claremont
Cause of Death: chronic myocarditis

Fannie Gill
12/21/1922
79
Admitted: 12-13-1920
Born in: Maine
Sending Town: Claremont
Cause of Death: shock

Mary Severance
1-21-23
84
Admitted: 11-8-1913
Single
Born in: Washington
Sending Town: Charlestown
Cause of Death: old age

John Haley
3-9-1924
93
Admitted: 1-12-1920
Born in: Canada
Sending Town: Claremont
Cause of Death: senility

Arthur Bailey 1858-1927
Admitted: 11-15-1912
Born in: Portland, Maine
Sending Town: Newport
Cause of Death: heart disease
Date of birth: 10-19-1858
Date of death: 1-17-1927

Row three

Sally Cheney
Admitted: 1/14/1884
Born in: Sunapee
Sending Town: Sunapee
Died: Feb. 22, 1888
Age 67

And
Liza Colby
Admitted: 9/22/1867
Born in: Grantham
Sending Town: Plainfield
Single

George W. Cary
5-23-1888
79

Laura Hibbard
5-9-1888
38
Admitted: 10-11-1876, age 36
Born in: Grantham
Sending Town: Grantham

Frank Davis
8-23-1888
71
Admitted: 10-1-1887
Born in Epsom, New Hampshire
Sending Town: Unity
Married

Alvin Sanborn
1-3-1889
72
Admitted: 2-16-1869 at age 55
Single
Born in: Springfield
Sending Town: Springfield

Joseph Cole
May 20, 1890
Age 64
Admitted: 11-17-1879
Born in: Canada
Sending Town: Springfield
Married

Mary Evans
6-1-1889
Admitted: 7-29-1887
Born in: Windsor, Vermont
Sending Town: Newport
Married
Report states that she died on 5-30-1889

Sally Harrington
1-9-1876
Admitted: 1-7-1868
Single
Born in: Plainfield
Sending Town: Plainfield

Abram Hainsworth
10-6-1889
80
Admitted: 4-19-1889
Born in: England
Sending Town: Newport
Married
According to the report he died on 10-3-1889

Jane Billings
12-26-1889
55
Admitted: 3-10-1874 at age 34
Born in: Ireland
Sending Town: Claremont
Widow

Abigail Goodwin
Single
Age 78
Admitted 12/30/1868
Born in: Charlestown
Sending Town: Charlestown

Laura Goodhue
2-21-1890
Age 7
Admitted: 8-18-1889
Born in: Newport
Sending Town: Croydon
Cause of Death: consumption
Report states she died 2-19-1890

Ransom Weed
2-26-1890
79
Admitted: 5-12-1889
Sending Town: Acworth
Single

Thomas Mulqueen
4-3-1890
Admitted: 12-21-1889
Sending Town: Charlestown
Cause of Death: consumption
Andrew Spaulding
Admitted: 3/26/1886
Sending Town: Plainfield
Born in: Holland
Age: 75
Married

Peter Benway
10-1-1890
70
Admitted: 7-20-1889
Sending Town: Claremont
Married
Report states he died 9-30-1890

Joel Chase
1-28-1892
55
Admitted: 1-24-1869
Born in: Croydon
Sending Town: Croydon
Single
Cause of Death: apoplexy

Hannah Weeks
2-6-1892
65
Admitted: 2-11-1889
Sending Town: Grantham
Married
Cause of Death: Bright's Disease

Moody Stiles
9-15-1892
91
Admitted: 5-16-1887
Born in: Greenfield, Massachusetts
Sending Town: Plainfield
Single
Cause of Death: old age
Book says died 9-12-1892

Merrill Kendall
6-25-1893
75
Admitted: 5-8-1891 at age 73
Born in: Washington
Sending Town: Washington
Married
Cause of Death: Bright's Disease

Emily Heywood
10-9-1893
69
Admitted: 8-14-1893
Sending Town: Newport
Cause of Death: Bright's Disease

Darius Chamberlin
12-7-1894
65
Admitted: 12-8-1892
Born in: Montpelier, Vermont
Sending Town: Newport
Cause of Death: heart disease

Rosilla Lang
1-29-1895
63
Admitted: 9-26-1868
Born in: Goshen
Sending Town: Goshen
Single
Cause of Death: Bright's Disease

Edwin Walker
3-31-1895
Admitted: 3-4-1895
Born in: Claremont
Sending Town: Claremont
Married
Cause of Death: Bright's Disease

Thomas Crooker
4-15-1895
Age 76
Admitted: 6-24-1893
Born in: Concord, New Hampshire
Sending Town: Claremont
Cause of Death: consumption

Marilla Locke
7-7-1897
95
Admitted: 2-27-1876
Born in: Gilsum
Sending Town: Lempster
Married

James Allen
12-25-1898
40
Admitted: 12-20-1898
Born in: Nashua
Sending Town: Claremont
Cause of Death: consumption

Warren Stearns
Jan 1-1899
71
Admitted: 12-26-1879 at age 56
Single
Born in: Plainfield
Sending Town: Claremont
Cause of Death: cancer of the stomach

William Campbell
9-8-1900
63
Admitted: 7-26-1900
Born in: Scotland
Sending Town: Grantham
Cause of Death: consumption

Fannie Wright
5-4-1901
69
Admitted: 7-21-1900 at age 68
Single
Born in: Randolph, Vermont
Sending Town: Claremont
Cause of Death: consumption

Hirman Barber
6-10-1901
40

Albert Spaulding
7-22-1901
43
Admitted: 12-28-1880 at age 24
Single
Born in: Reading, Vermont
Sending Town: Cornish
Cause of Death: sunstroke

Row 4

Susan Fuller
6-24-1878
87
Admitted: 11-24-1877 at age 86
Born in: Maine
Sending Town: Springfield
Cause of Death: old age

Lucy Lewis
7-9-1878
70
Admitted: 11-15-1875
Sending Town: Claremont

Betty Carr
7-13-1878
110
Admitted: 9-18-1875
Born in: Salisbury, Massachusetts
Sending Town: Claremont
Single

Melinda Dunbar
5-28-1879
70
Admitted twice: at age 58 and again at age 70
Born in: Grantham
Sending Town: Grantham
Single

Drucilla Davis
8-1-1878
58
Admitted: 10-11-1872 at age 52
Born in: Sunapee
Sending Town: Sunapee

William
October 15, 1921
Stone is too decayed to read

Levi Nelson
9-17-1879
68
Admitted: 1-13-1877
Born in: Portsmouth, New Hampshire
Sending Town; Newport
Married

Nancy Belanger
12-11-1879
89

Prudence B. Blood
8-25-1876
Admitted: 5-29-1873
Born in: Charlestown
Sending Town: Charlestown
Single

Rhoderick Alexander
2-23-1884
78
Admitted: 10-31-1814
Sending Town: Plainfield
Cause of Death: pneumonia

James Broley
8-28-1880
Admitted: 11-29-1879
Born in: Liverpool, England
Sending Town: Merrimack County
Married

Mary Ann Brown
4-13-1883
71
Admitted: 1-7-1868
Born in: Canada West
Sending Town: Washington

Henry Picknell
12-9-1885
62

Harper Gault
3-11-1886
77

Nathaniel Penniman
3-29-1886
Age 70
Widower
Admitted: 3-21-1876
Sending Town: Claremont
Born in: Windsor, Vermont

Joseph Chambers
11-30-1886
63

Infant of D& E Jordan
8-8-1886

John Jordan
9-9-1886
2

George Stone
10-10-1886
78
Admitted: 8/13/1881
Sending Town: Cornish
Born in: Danville, Vermont
Age: 63

George…dane
10-23-1881
73
Stone is decayed

William West
12-22-1886
68

Thomas Anderson
3-8-1887
68

John Blood
4-21-1887
65
Book gives date of death as 4-19-1887

John Van Goskey
6-16-1887
30

Moses Atwood
8-18-1893
77
Admitted: 4-30-1884 at age 66
Born in: Alexandrer
Sending Town: Croydon
Single

Louis Gentle
7-8-1898
33

Jason Darling
9-15-1898
32
Admitted: 12-13-1984 at age 28
Born in: New York City
Sending Town: Newport
Single
One of only two African American residents in the cemetary

Joseph Martelle
10-24-1898
51
Admitted: 8-1-1896
Born in: New York
Sending Town: Claremont

Asa Way
February 16, 1890
Admitted: 9-9-1889 age 82
Sending Town: Lempster

William Brownell
8-31-1901
85
Admitted: 10-16-1899
Born in: Pomfret, Vermont
Sending Town: Springfield

Daniel Davis
5-2-1901
82
Admitted: 2-28-1894
Born in: Bridgewater, Vermont
Sending Town: Springfield
Cause of Death: Bright's Disease
Married

James Sanders
1888-1940
Born in: Ticonderoga, New York
Sending Town: Claremont
Widower and disabled

Row 5

Charles Day
3-13-1870
78
Admitted: 8-24-1868
Single
Born in: Hillingby, Connecticut
Sending Town: Croydon

Joseph Garfield
4-8-1869
61
Admitted: 8-18-1869
Born in: Langdon
Sending Town: Langdon

John Draper
12-1-1869
86
Admitted: 12-3-1868
Born in: Claremont
Sending Town: Claremont
Single
Cause of death: softening of the brain

William Brink
10-30-1869
68
Born in: Lebanon, New Hampshire
Sending Town: Plainfield
Single

John Hunt
6-1-1868
83
Admitted: 3-12-1868
Born in: Strafford, England
Sending Town: Plainfield
Cause of Death: apoplexy
Medical note: Was well in the morning

Benjamin Edminster
1-8-1868
73
Admitted: 12-16-1867 (the first day the almshouse was open)
Sending Town: Cornish
Cause of Death: tertiary syphilis

Nancy Fletcher
9-8-1873
67
Born in Salisbury, New Hampshire
Sending Town: Springfield
Cause of Death: cancer of the tongue
Married

Allen Bundy
12-26-1873
70
Born in: Middletown, Connecticut
Sending Town: Charlestown
Cause of Death: consumption
Widower

George Linton
August 1, 1876
63

Amanda Field
3-11-1877
75
Admitted: 10-18-1871
Born in: Claremont
Sending Town: Newport
Cause of Death: dropsy
Single

Joseph Morris
10-9-1877
61
Admitted: 3-2-1876
Born in: Canada
Sending Town: Claremont
Cause of Death: kidney disease
Married

Willis Blackwell
10-29-1877
Admitted: 12-30-1876
Born in: North Carolina
Sending Town: Claremont
Single
Black

Charles Lucy
1877
42
Admitted: 9-22-1868
Born in: Cornish
Sending Town: Cornish
Cause of Death: paralysis

John Whelen
9-9-1883
67
Admitted: 7-10-1877
Born in: Lempster
Sending Town: Lempster
Cause of Death: old age

Charles Howard
1-10-1878
78
Admitted: 12-5-1870
Born in: Richmond, Massachusetts
Sending Town: Charlestown
Single

Lewis Hosdgkins
5-22-1878
86
Admitted: 2-1-1872
Born in: Langdon
Sending Town: Charlestown
Single

Mary Barber
4-2-1878
88
Admitted: 11-24-1887
Born in Salisbury, New Hampshire
Sending Town: Springfield
Cause of Death: old age

George Smith
9-12-1880
72
Admitted: 10-14-1875
Born in: Langdon
Sending Town: Langdon
Cause of Death: Dropsy
Single

Mary Buckley
10-25-1880
67

Rhoda Russell
1883
Admitted: 10-10-1876
Born in: Springfield
Sending Town: Springfield
Single

Milan Chatterton
4-2-1883
60

Isaac Kirby
11-10-1883
78
Admitted: 10-20-1876
Born in: Ireland
Sending Town: Lempster
Single
Cause of Death: pneumonia

Sarah Bradbury
11-15-1883
81
Admitted: 10-18-1867
Born in: Unity
Sending Town: Unity
Cause of Death: dropsy

Camilla Waterhouse
5-14-1898
71

Doris Jones
5-30-1898
2 weeks
Admitted: 5-15-1898
Born in: Langdon
Sending Town: Langdon
Cause of death: shovela jan fantrum (failure to thrive)

Aaron Hand
3-7-1899
Age 76
Born in: Bath, New Hampshire
Sending Town; Croydon
Cause of Death: bronchitis
Married

Calvin Clemons
6-3-1901
59
Admitted: 2-7-1898
Born in: Wallingford, Connecticut
Cause of Death: paralytic shock
Married

Row 6

John Collins
10-17-1872
50
Admitted: 4-25-1872
Born in: Springfield
Sending Town: Springfield
Single

Susan Davis
4-21-1872
87
Admitted: 3-14-1868
Born in: Sutton, New Hampshire
Sending Town: Plainfield

Amos Parker
11-4-1869
84
Admitted: 12-21-1867
Born in: Charlestown
Sending Town: Charlestown
Cause of death: old age

George Kinsgbury
8-13-1872
77
Admitted: 1-19-1869
Born in: Alstead, New Hampshire
Sending Town: Langdon

Amos Buckman
7-6-1871
85
Admitted: 6-17-1871
Born in: Unity
Sending Town: Unity
Cause of Death: scrapula/consumption
brought in from Springfield, Vermont by CJ Rumrill and
married to:

Betsy Buckman
2-22-1873
85
Admitted: 6-17-1883
Born in: Unity
Sending Town: Unity
Cause of Death: consumption
Widow

Lucy Davis
4-3-1873
78
Admitted: 12-16-1867
Born in: Hartland, Vermont
Sending Town: Newport
Cause of Death: consumption

Mary Cooper
6-4-1873
51
Admitted: 11-8-1871
Born in: Sunapee
Sending Town: Newport
Cause of Death: consumption
Widow

Philandra Locke
6-1-1884
71
Admitted: 2-28-1882
Born in: Sharon, Vermont
Sending Town: Charlestown
Widow

Elias Harrington
6-21-1873
84
Admitted: 1-7-1868
Born in: Plainfield
Sending Town: Plainfield
Cause of Death: dropsy
Single

Lydia Livermore
12-26-1873
73
Born in: Windsor, Vermont
Sending Town: Washington
Widow

Jonas Weeks
5-26-1876
72
Admitted: 1-27-1868
Born in: Long Island, New Your
Sending Town: Unity

Abigail Wheeler
9-1-1876
Admitted: 12-30-1867
Born in: Charlestown
Sending Town: Charlestown

Anna Wallace
7-7-1877
79
Admitted: 12-16-1867
Born in: Merrimack, New Hampshire
Sending Town: Newport
Cause of Death: old age

Noah Porter
2-6-1876
71
Admitted: 8-6-1867
Born in: Charlestown
Sending Town: Charlestown

Daniel Heath
8-16-1877
Admitted: 5-13-1877
Born in: Bangor: Maine
Sending Town: Charlestown
Cause of Death: "died of a very bad disease in connection
with delirium tremors"

John Bell
5-1-1877
71
Admitted: 1-16-1871
Born in :Greenfield, Massachusetts
Sending Town: Cornish
Cause of Death: dropsy
Married

John Wheeler
9-3-1877
81
Admitted: 7-10-1877
Born in: Lempster
Sending Town: Lempster
Cause of Death: old age

Charlotte Abbott
9-18-1877
Age 79
Admitted: 2-28-1871
Born in: Manchester, New Hampshire
Sending Town: Charlestown
Widow

Mary Kenyon
2-11-1871
86
Admitted: 7-7-1868
Born in: Plainfield
Sending Town: Plainfield
Single

Sybill Kendall
3-26-1878
75
Admitted: 9-16-1867
Born in: Swanzey, New Hampshire
Sending Town: Claremont
Cause of Death: liver cancer
Single

Daniel Hall
6-22-1878
90
Admitted: 10-4-1875
Born in: Bowe, New Hampshire
Sending Town: Sunapee
Cause of Death: old age
Married

Deborah Hall
5-7-1880
75
Admitted: 10-17-1875
Born in: Hooksett, New Hampshire
Sending Town: Sunapee
Cause of Death: consumption
Married

Charles Duciley
12-14-1881
87

Henry Jones
5-11-1883
73
Admitted: 9-12-1877
Born in: Stoddard, New Hampshire
Sending Town: Goshen
Single

Mariah Whittaker
11-19-1883
68
Admitted: 12-14-1876
Born in: Goshen
Sending Town: Unity
Married
Cause of Seath: pneumonia

Little Mary Lovering
8-23-1883
Admitted: 5-1-1880
Born in: Croydon
Sending Town: Croydon
Single

Mary Lovering
8-15-1883
Born in: Croydon
Sending Town: Croydon
Married
Cause of Death: consumption

Robert Phelps
8-21-1883
Born: Almshouse 10-13-1882

Martha Wilson
1898
Age 52
Admitted: 2-11-1898
Born in: Canada
Sending Town: Lempster
Cause of Death: senile gangrene

John Cutts
5-18-1899
2
Born: County Almshouse 3-12-1897
Died from scalding

Gustave Knowland
2-23-1900
40
Admitted: 9-22-1899
Born in: Finland
Sending Town: Croydon
Cause of Death: paresis
Single

Known to be buried at the county cemetery but no stone found:

Joseph Welch 3-14-1922 stillborn

Lombard stillborn 7-6-1929 mother Louise Lombard

Linus Dickerson 3-22-1897 age 71

John Bushway admitted 5-4-1901 age 70 born in Canada, sent from Langdon married died 11-8-1901 paralysis

Frances Clemons admitted 2-7-1898 born in Charlestown sent from Newport died 12-17-1898 from pneumonia

Emma Wilson admitted 2-11-1898, born in Canada, sent from Lempster, married, senile gangrene, age 52, died 4-28-1898

Edna Nichols admitted 6-4-1895 born in Boston, Mass., sent to the Concord asylum, single died 9-6-0904 from paresis

Edward Hall admitted 10-29-1899, born in Thornton, New Hampshire, sending town: Unity, cause of death: old age, age 87

Eliza Forehand, admitted 6-12-1882; born: Montpelier, Vermont, sending town: Cornish, widow, cause of death: pneumonia

Robert Phelps, born at the Almshouse 10-13-1882